Together in our World

PROTECTING THE PLANET

Louise Spilsbury
Illustrated by Hanane Kai

Published in North America by Free Spirit Publishing Inc., Minneapolis, Minnesota, 2020

North American rights reserved under International and Pan-American Copyright Conventions. Unless otherwise noted, no part of this book may be reproduced, stored in a retrieval system, or transmitted in any form or by any means, electronic, mechanical, photocopying, or otherwise, without express written permission of the publisher, except for brief quotations or critical reviews. For more information, go to freespirit.com/permissions.

Free Spirit, Free Spirit Publishing, and associated logos are trademarks and/or registered trademarks of Free Spirit Publishing Inc. A complete listing of our logos and trademarks is available at freespirit.com.

Library of Congress Cataloging in Publication Data
This book has been filed with the Library of Congress.
LCCN: 2020933288
ISBN-13: 978-1-63198-577-5

Free Spirit Publishing does not have control over or assume responsibility for author or third-party websites and their content. At the time of this book's publication, all facts and figures cited within are the most current available. All telephone numbers, addresses, and website URLs are accurate and active; all publications, organizations, websites, and other resources exist as described in this book; and all have been verified as of January 2020. If you find an error or believe that a resource listed here is not as described, please contact Free Spirit Publishing. Parents, teachers, and other adults: we strongly urge you to monitor children's use of the internet.

Reading Level Grade 3; Interest Level Ages 6–10
Fountas & Pinnell Guided Reading Level N

10 9 8 7 6 5 4 3 2 1
Printed in China
H13770320

Free Spirit Publishing Inc.
6325 Sandburg Road, Suite 100
Minneapolis, MN 55427-3674
(612) 338-2068
help4kids@freespirit.com
freespirit.com

First published in Great Britain in 2020 by Wayland, an imprint of Hachette Children's Group

Text copyright © Hodder and Stoughton, 2020. Illustrations copyright © Hanane Kai, 2020. All rights reserved.

The rights of Hanane Kai to be identified as the illustrator of the Work have been asserted by her in accordance with the Copyright, Designs and Patents Act, 1988.

Edited by Sarah Peutrill. Designed by Hanane Kai. Texturing of illustrations by Mariela Gallegos.

Free Spirit offers competitive pricing.
Contact edsales@freespirit.com for pricing information on multiple quantity purchases.

Contents

Planet Earth — 2

Making a Mess — 4

Pollution — 6

Global Warming — 8

Too Much Stuff! — 10

Forests Under Threat — 12

A World of Waste — 14

People Who Help — 16

Reduce Waste — 18

Reuse — 20

Recycle — 22

Use Less Power — 24

Enjoy Our Planet! — 26

Find Out More — 28

Glossary — 29

The planet we live on is full of animals, from tiny insects to huge whales—and us! Our amazing Earth gives us and other animals everything we need to live.

Our planet provides air to breathe, water to drink, and food to eat. We use plants and other natural resources to make clothing, homes, and medicines. We use fuels from the ground to power machines. Planet Earth helps us, so we need to protect it!

When our planet is clean and healthy, the people, plants, and animals on it can be clean and healthy too. When we damage parts of the planet, we spoil it for all living things.

Think about a lovely park. Then imagine it filled with litter. It looks ugly. Animals looking for food get trapped in cans or hurt by sharp edges. A park that is dirty and filled with waste is not fun or safe.

Our planet is being spoiled by waste. A whole truckload of plastic waste ends up in oceans every single minute! Anything that makes the planet dirty and unhealthy is called pollution.

Air pollution happens when people burn fuel to power cars or factory machines. Burning fuel releases smoke and gases into the air. Breathing this polluted air can make people sick.

Extra gases are released into the air when we burn fuel. These gases cause another problem. They wrap around the planet like a blanket and trap the sun's warmth. They make Earth hotter. This is known as global warming.

Global warming causes big problems. Too much heat can make people sick. Extra heat can start fires that burn forests and homes. When there is a lot of heat and not enough rain, plants die and people and animals go hungry.

We use a lot of electricity, fuel, and natural resources to make new things. To make a plastic toy, we take oil from the ground to make the plastic. We dig up coal to make electricity to power factory machines that make the toys. Trucks use lots of fuel taking the toys to stores.

People are making and buying so much new stuff that the planet's natural resources are being used up. Making new things causes more pollution and global warming.

Another natural resource we are using up too quickly is trees. People cut down trees to make paper and furniture. We cut down whole forests to clear land to build cities, raise farm animals, and grow food. About 27 soccer fields' worth of trees are lost every minute!

We need trees. They help keep the planet's air healthy. Trees make oxygen, which is the main gas in the air we breathe. Trees also clean the air by taking in some of the gases that cause air pollution and global warming.

People throw away a lot of unwanted stuff when it gets broken or old. Some of this waste, like plastic, is very bad for the planet. Plastic doesn't rot away like paper or food. It can hang around on the planet for hundreds of years.

When plastic, food, and other waste, like old computers, break down in landfills, it causes big problems. Trash releases some of the gases that make Earth hotter. It also leaks substances into the ground that pollute the water living things need to survive.

The good news is that lots of people are working to protect our planet. Some countries have laws to stop people from cutting down trees. Others have laws to stop factories and power stations from making pollution.

There are laws to stop stores from giving out plastic bags that are used only once before being thrown away. Some countries have laws that punish people for dropping litter.

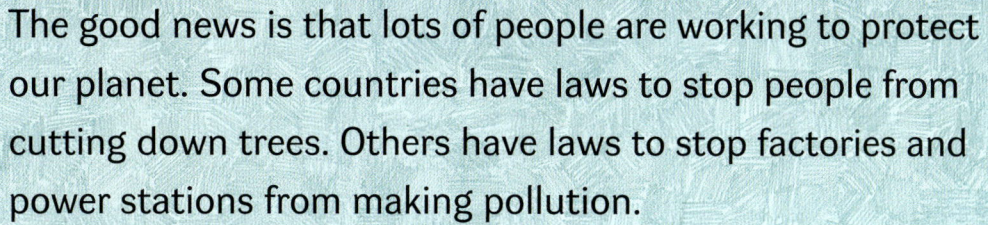

We can all help protect the planet. We can reduce the amount of stuff we buy and throw away. We can borrow things from a friend or buy them from a secondhand store instead.

To reduce waste, we can take our lunch in a reusable lunchbox or bag instead of a plastic bag. And we can bring drinks in a bottle that can be reused, rather than buying drinks that come in cans or plastic bottles that we use only once.

We can reuse lots of things instead of throwing them away. We can fix things when they break so that we can use them again. Broken toys can be glued back together. Holes in jeans can be mended.

It's fun to reuse things in new ways. Why not use empty packaging for art projects or to store things like colored pencils? Instead of throwing out old and odd socks, you could make them into funny sock puppets! What else can you reuse?

21

Lots of things can be recycled and remade into new products. This saves electricity, fuel, and natural resources. We can recycle glass, plastic bottles, cans, and even old shoes and clothes.

At a recycling center, cans, bottles, and paper are crushed, broken down, and later turned into new cans, bottles, and paper. Plastic bottles can also be recycled into brand-new things like fleece jackets and kayaks. It takes about 10 bottles to make a new T-shirt!

You can also try to use less fuel and electricity. Turn off the lights when you leave a room. Put on a sweater to keep warm instead of turning on a heater. Take a short shower instead of a deep bath. Showers use less water and less energy to heat the water.

Sharing rides saves fuel too. You could travel by bus or get a lift with a friend to school or a club. Better still, walk or ride a bike instead. This saves fuel and gives you some exercise!

Another way to care for our big, beautiful planet is to get outside and enjoy it! You could go for a walk, visit a local park, or spend time at a nature center. You could paddle in a stream or explore a tide pool. What's your favorite thing to do outdoors?

We need to look after our planet, because our planet looks after us. If we all make small changes to protect the planet, we will see big results.

Find Out More

Books

Go Green!
Liz Gogerly, Free Spirit Publishing, 2019

Let's Walk to School (Good to be Green)
Deborah Chancellor, Wayland, 2019

Make a Splash!
Cathryn Berger Kaye and Philippe Cousteau, with EarthEcho International, Free Spirit Publishing, 2013

What a Waste
Jess French, DK Publishing, 2019

Websites

You can learn about the natural world and ways to protect it at PBS Kids Cyberchase: pbskids.org/cyberchase/topics/environment

Learn about Earth's oceans and how to protect them at National Ocean Service Just for Kids: oceanservice.noaa.gov/kids

You can learn more about the planet and its problems at: climatekids.nasa.gov

Glossary

coal: a black rock that can be burned to release energy in the form of heat

fuel: a substance such as coal or oil that can be burned to produce heat or power

gas: a substance that is like air and has no fixed shape. Some gases can be burned as fuel

global warming: the rise in Earth's temperature, probably caused by human activities such as burning fuel

landfill: a place where things that are no longer wanted are taken

law: a rule that is set by the rulers of a country

natural resource: something that is found in nature and can be used by people

oil: thick black liquid that can be used as fuel to power engines

pollution: something that makes air, soil, or water dirty and unhealthy

power station: a factory where electricity is made

secondhand store: a place selling used items

More Great Books from Free Spirit

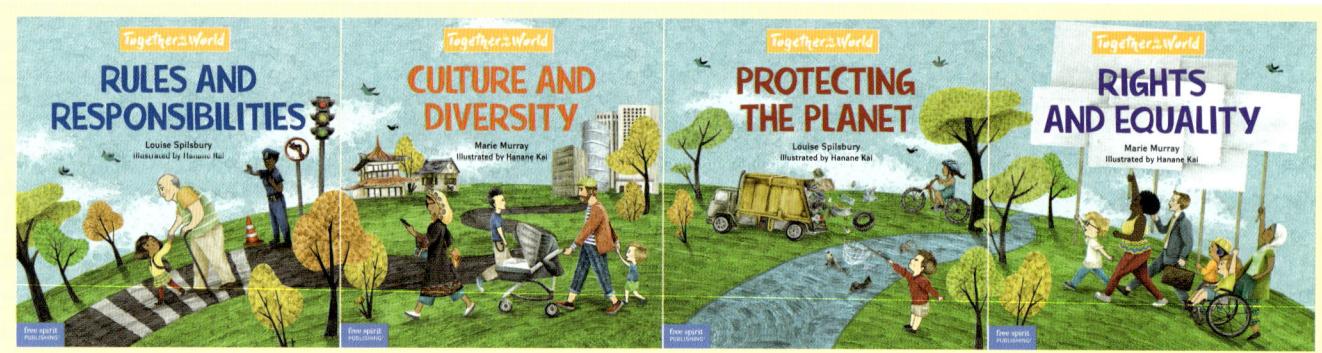

Together in Our World Series
The Together in Our World series addresses issues children are likely to hear about in the news and may not understand. In a straightforward and kid-friendly way, these picture books explain tough topics and offer readers ideas for what they can do to help make the world a better, fairer place.

Each book: 32 pp.; full-color; HC; 8½" x 8½"; ages 6–10

Go Green!
Join the green team and learn how to reduce, reuse, and recycle!
by Liz Gogerly, illustrated by Miguel Sanchez
48 pp.; HC; full-color; 8¼" x 10½"; ages 8–12.

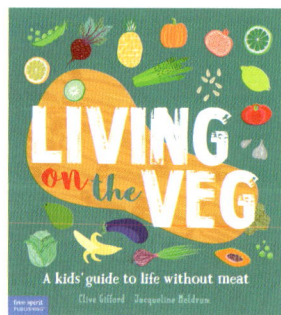

Living on the Veg
A kids' guide to life without meat
by Clive Gifford and Jacqueline Meldrum
80 pp.; HC; full-color; 8¼" x 9"; ages 8–13.

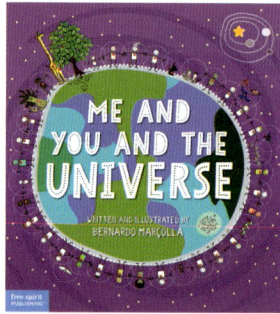

Me and You and the Universe
written and illustrated by Bernardo Marçolla
36 pp.; HC w/ jacket; full-color; 8¼" x 9"; ages 3–8.

Interested in purchasing multiple quantities and receiving volume discounts?
Contact edsales@freespirit.com or call 1.800.735.7323 and ask for Education Sales.

Many Free Spirit authors are available for speaking engagements, workshops, and keynotes.
Contact speakers@freespirit.com or call 1.800.735.7323.

For pricing information, to place an order, or to request a free catalog, contact:

Free Spirit Publishing Inc. • 6325 Sandburg Road • Suite 100 • Minneapolis, MN 55427-3674
toll-free 800.735.7323 • local 612.338.2068 • fax 612.337.5050 • help4kids@freespirit.com • freespirit.com